RADICALS IN CHEMISTRY, PAST AND PRESENT

THE CHANDLER LECTURE

1928

COLUMBIA UNIVERSITY PRESS
COLUMBIA UNIVERSITY
NEW YORK

FOREIGN AGENTS
HUMPHREY MILFORD
AMEN HOUSE, E. C.
LONDON

EDWARD EVANS AND SONS, LTD.
30 N. SZECHUEN ROAD
SHANGHAI

RADICALS IN CHEMISTRY, PAST AND PRESENT

BY

MOSES GOMBERG

University of Michigan, Ann Arbor, Michigan

New York
COLUMBIA UNIVERSITY PRESS
1928

All rights reserved

Copyright, 1928
By COLUMBIA UNIVERSITY PRESS

Printed from type, February, 1928

Printed by Mack Printing Co.
Easton, Pa., U. S. A.

Moses Gomberg

Radicals in Chemistry, Past and Present

Moses Gomberg
UNIVERSITY OF MICHIGAN, ANN ARBOR, MICH.

The following Chandler Lecture at Columbia University was delivered in Havemeyer Hall on the evening of December 16, 1927, in accordance with the provisions of the Charles Frederick Chandler Foundation and on the occasion of the presentation of the Chandler gold medal, which was awarded to Professor Moses Gomberg by the Trustees of Columbia University. This was the tenth award and was made to Professor Gomberg for his outstanding work on Tri-valent Carbon and on Free Radicals."

THE conditions that go with the nomination of the speaker for this happy occasion extend to him the privilege, and in a measure make it incumbent upon him, to present a brief account of his own contribution in some branch of chemistry. An attempt to give such an account in this instance is not without difficulties, largely because of the specialized nature of the topic to be discussed. None the less, a brief survey of the history of free radicals will be given, with the emphasis mainly on those aspects which may not be devoid of general interest.

RADICALS 1815-58[1]
Lavoisier's Radicals (1785-1815)

The term "radical" we owe to Lavoisier. In accordance with his conception of the supreme role played by oxygen, to Lavoisier the great majority of the mineral substances were oxygen compounds, the bases being oxidized metals, the acids oxidized non-metals, and the salts combinations of these two. The fundamental stuff which through its oxidation gives rise to base or acid Lavoisier designated as the "radical." At that time (1785) about twenty-five of these

[1] The writer feels particularly indebted to Lowry's "Historical Introduction to Chemistry," Macmillan and Co. (1915); Hjelt's "Geschichte der organischen Chemie," Vieweg und Sohn (1916); Graebe's "Geschichte der organischen Chemie," J. Springer (1920); Lipmann's "Zeittafeln zur Geschichte der organischen Chemie," J. Springer (1921).

radicals—elements as we call them now—were known to be capable of existence in the free state. Lavoisier, however, had no doubt that in time still other such radicals would be isolated from their acid or base combinations.

We owe to Lavoisier also the first methods of determining the composition of so-called organic compounds. He analyzed alcohol, olive oil, tallow, fruit acids, etc., and his conclusion was this:

> I have observed that in the mineral kingdom almost all oxidizable radicals were *simple;* that in the vegetable kingdom, on the contrary, and above all in the animal kingdom, there were scarcely any which were composed of less than two substances, hydrogen and carbon; that often nitrogen and phosphorus were also present; in this manner radicals were produced consisting of four components.

Very little progress was made during the next fifteen years. Nor did the first decade of the nineteenth century bring any important advance in the field of organic chemistry, rich as that decade proved for inorganic chemistry: Berthollet's "Essai de statique chimique" (1803), Proust's Law of Constant Proportions (1801–8), Dalton's Law of Multiple Proportions and the beginnings of his Atomic Theory (1804), the electrolytic separation of the alkali metals by Davy (1807), Gay-Lussac's Law of Combining Volumes (1808), Berzelius' first paper on the "Constancy of Composition in the Inorganic Compounds (1810), and finally Avagadro's Hypothesis (1811). The principal reason why organic chemistry remained nearly stationary for the twenty-five years following the work of Lavoisier was that no satisfactory and ready method for the quantitative analysis of organic substances was known. There was no way of testing old or new theories. This became possible only after Gay-Lussac and Thenard, in 1810, presented a fairly accurate method of determining quantitatively the composition of organic substances. Berzelius soon introduced further important improvements, so that in 1813, with his wonderful skill in analyzing, he could make an analysis in three months and was able to carry through in one year the analysis of as many as fourteen substances.

The Cyanogen Radical

The new methods of analysis soon began to yield fruit. The year 1815 marks indeed a milestone in the history of organic chemistry. Gay-Lussac prepared for the first time anhydrous, pure prussic acid, and determined its composition as HCN. He made from the acid a number of salts and other derivatives and found that all of them retained the part (CN) originally present in the acid. Said Gay-Lussac:

> There is, then, a very great analogy between prussic acid and hydrochloric and hydriodic acid: like them, it has a radical which combines with potassium, and forms a compound quite analogous with chloride and iodide of potassium: only this radical is *compound*, where chlorine and iodine are *simple*.

Gay-Lussac designated the radical of prussic acid as "cyanogen," and finally he succeeded in obtaining the cyanogen gas itself in the free, uncombined state. For the first time a compound radical had been isolated as such. Lavoisier's prophecy had been vindicated, and his early generalizations could now be reaffirmed with greater force. Said Berzelius in 1817:

> In inorganic nature all oxidized bodies contain a *simple radical*, while all organic substances are oxides of *compound radicals*. The radicals of vegetable substances consist generally of carbon and hydrogen, and those of animal substances of carbon, hydrogen and nitrogen.

Henceforth this dualistic theory of organic matter—i. e., radical plus oxygen—persisted in its various guises for almost fifty years, and the theory of compound radicals played in one form or another a dominant part in the development of organic chemistry. Are all the vegetable acids derived from one and the same compound radical? If not, how many different such radicals are there? What is the radical of the carbohydrates, of fats? What is the exact composition of each individual radical? May not the molecule contain several radicals, all equally important? Does the radical really represent a compact cluster which is as immutable and as indivisible as are the elementary atoms in the mineral kingdom? Or are these radicals themselves subject to change? If changeable, why consider them as of greater importance in the molecule than the other components?

One after another these questions came to the fore, and the answer to each brought with it a slightly different hypothesis in regard to the constitution of organic molecules.

The Etherin Theory (1827-37)

As far back as 1815 Gay-Lussac, on the basis of the respective vapor densities of alcohol and ether, arrived at the conclusion that alcohol might be a combination of one volume of ethylene with one of water, and ether of two volumes ethylene with one of water. About a dozen years later Dumas and Boullay determined carefully the composition of alcohol and of ether, and they confirmed the conclusions of Gay-Lussac. Dumas and Boullay also determined the composition of various organic esters of alcohol, and made the sweeping suggestion that alcohol, ether, and their esters could all be considered as derivatives of ethylene. In justification of their view they pointed to these facts: (1) ethylene is actually obtainable from alcohol; (2) ethylene combines with sulfuric acid, and from this combination alcohol or ether can be regenerated, as had been shown by Faraday and Hennel. The evidence seemed complete that ethylene was the compound radical which preëxisted in all these substances:

Substance	Modern Formula	Dumas' Formula
Ethylene	C_2H_4	C_2H_4(etherin)
Alcohol	C_2H_6O	$C_2H_4 \cdot H_2O$
Ether	$(C_2H_5)_2O$	$C_2H_4 \cdot \frac{1}{2}H_2O$
Ethyl chloride	C_2H_5Cl	$C_2H_4 \cdot HCl$
Ethyl acetate	$C_2H_5 \cdot C_2H_3O_2$	$C_2H_4 \cdot C_2H_4O_2$

The etherin thus constitutes historically the second compound radical for the independent existence of which claim was being made. Later the etherin theory was extended to methyl alcohol and to other alcohols as well. For the first time it became possible to classify systematically organic compounds into groups, each group around one common radical—ethylene, methylene, etc. At first Berzelius accepted the etherin theory, but later he began to doubt the validity of this view, because Dumas' formulas for the derivatives of etherin did not appear to Berzelius, nor to Liebig, to harmonize with their behavior. If alcohol and ether were really hydrates, then calcium oxide should convert them into ethylene—and yet it did not. The etherin theory, none the less, held sway for ten years, when Dumas finally conceded that ethyl, rather than ethylene, was likely to be the true compound radical. Thus etherin, as a radical with independent existence, disappears from the stage, giving place to the hypothetical ethyl of Berzelius and Liebig.

Benzoyl Radical

Of unusually great influence proved to be the classic investigation of Wöhler and Liebig on the benzoic acid radical.

This extensive study was commenced in the middle of May, 1832, and was completed in August of the same year. Starting with oil of bitter almonds, our benzaldehyde, they found that they could convert it successively into a series of related compounds, in all of which the group (C_7H_5O) remained unaltered:

Oil of bitter almonds	$C_7H_5O.H$
Benzoyl chloride	$C_7H_5O.Cl$
Benzoyl cyanide	$C_7H_5O.CN$
Benzoic acid	$C_7H_5O.OH$
Benzamide	$C_7H_5O.NH_2$
(Benzoyl)	($C_7H_5O.$)

To quote from them:

Organic substances may be arranged around a common center, a group of atoms preserving intact its nature and composition amid the most varied associations with other components. This stability, this analogy, justifies one to regard this group as equivalent to an element, as a kind of a compound element; we propose for this radical the special name Benzoyl.

Wöhler and Liebig's investigation, and its exceptionally favorable reception by Berzelius, produced a strong impression, notwithstanding the fact that the benzoyl radical itself seemed to elude isolation in the free state. Within two years after the publication of this famous investigation, there appeared a paper by Aug. Laurent,[2] in which the claim was made, with some slight reservations, that the very benzoyl radical itself was prepared by him. What an ironical incident! Laurent, who was destined to become in later years one of the bitterest opponents of the radical theory, should at this time, 1835, seem to be supplying Wöhler and Liebig with the very cornerstone of their benzoyl theory. More curious still is the fact that Liebig should be producing seemingly effective experimental evidence that Laurent's benzoyl has no connection with the true surmised radical of the benzoyl series.[3] And yet, from our present point of view, Laurent's substance was fully as much entitled to be classed among the radicals as any of the other compounds which, in subsequent years, were designated as free radicals, as it really was dibenzoyl ($C_6H_5CO)_2$.

The scheme proposed by Wöhler and Liebig of grouping a number of related substances around the benzoyl radical was soon extended by other investigators to other com-

[2] *Ann. chim.*, 59, 397 (1835).
[3] Hofmann, "The Life-Work of Liebig," Macmillan and Co., p. 82 (1876).

pounds. Piria, in Pisa, applied the theory to the salicylic acid group with its corresponding salicyl radical.

Radicals Are Not Immutable

It was becoming evident at this period that the groups which had been assumed to be the compound radicals were themselves not at all immune against change in composition when subjected to the influence of certain reagents. As argument against the immutability of radicals can be cited Regnault's investigation, which was carried out, strangely enough, in Liebig's own laboratory (1834-5). According to the etherin theory, ethylene dichloride was one of the most typical etherin compounds ($C_2H_4 + Cl_2$). Regnault found, however, that this substance, when heated with alcoholic potash, broke up into two substances as follows:

$$C_2H_4.Cl_2 = C_2H_3Cl + HCl$$

Obviously, the two chlorine atoms in the original compound played different functions—one atom being removable, the other not; consequently, not ($C_2H_4 + Cl_2$) is the make-up of the molecule, but ($C_2H_3Cl + HCl$). One of the two inevitable conclusions must now be drawn: either the compound radical (C_2H_4) is not immutable, or not (C_2H_4), but (C_2H_3), is the true radical. The defenders of the radical theory promptly adopted the second alternative. Liebig thought that this conclusion disposed once and for all of Dumas' etherin radical. He was right in that, but the conclusion worked equal havoc with his own ethyl radical—neither (C_2H_4) nor (C_2H_5) could now be considered as the unchangeable compound radical. Indeed, fifteen years previously Faraday obtained by the action of chlorine gas on $C_2H_4Cl_2$ the compound C_2Cl_6; consequently the group (C_2H_3), in its turn, cannot be considered the true radical. Then came the crucial experiments of Dumas and of Laurent, which finally culminated, by 1840, in their wellknown Theory of Substitution: Any hydrogen in the molecule, be it a part of the surmised radical or not, is equally subject to substitution by halogen, without very material change in the properties of the substance itself; for instance, compare acetic with trichloroacetic acid. The inference came about naturally that, not the composition of the radical in the molecule, but rather the type of the structure of the whole molecule was the responsible factor of the properties of the substance.

At first Liebig, as well as Berzelius, strenuously opposed the notion that radicals are subject to change. Moreover, as a follower of Lavoisier, Berzelius consistently rejected the inclusion of the electronegative oxygen or chlorine as an integral part of the radicals, and he was thus led to devise, with frequent revisions, new radicals for various substituted halogen compounds; for instance, benzaldehyde, previously considered as $(C_7H_5O).H$, became now $(C_7H_4).O$; tetrachloroether $(C_4H_6Cl_4O)$ became a conjugated compound $(C_2H_6O + C_2Cl_4)$. Liebig, on the other hand, saw the advantages of the substitution theory, but he still favored the theory of radicals in its general features for purposes of classification.

Do Radicals Have an Objective Reality? (1834–42)

Still another factor militated seriously against the theory of radicals. In 1834, in discussing the relation of ethylene, alcohol, and ethyl ether, Liebig used these words: "I doubt not that the radical of ether, namely, the hydrocarbon Ethyl, will be isolated in the free state." Several years later, however, the existence of ethyl as well as of other radicals was still only a matter of faith and of hope, Gay-Lussac's cyanogen and Dumas' doubtful etherin being the only two examples. Said Liebig in 1838:

> The organic radicals are known to us for the most part merely as hypothetical substances, but as regards their actual existence, there can be no more doubt than there is any doubt in regard to the existence of the inorganic radical of nitric acid (N_2O_5), although this, like ethyl, is as yet unknown.

It makes rather strange reading to find that Dumas, under the persuasive powers of Liebig, threw at this time (1837) his influence on the side of the radical theory, for a while at least. Jointly the two addressed to the Paris Academy a note entitled, "The Present State of Organic Chemistry." They made a passionate appeal in behalf of the radical theory, which they called the "natural classification of organic compounds."

> Organic chemistry possesses its own elements, some of which play a role similar to that of chlorine or oxygen in inorganic chemistry, and some play the role of metals.—Cyanogen, Amido, Benzoyl, are the very elements of which organic chemistry makes use in its operations.

Rhetoric alone was, however, not sufficient to counteract the influence of newer theories, based as these were upon the experimental evidence of substitution. If organic chemistry was to be defined with Liebig as the chemistry of radicals, "then," in the words of Gerhardt, himself a former pupil of Liebig, "it is the Chemistry of the Things Non-existent." Another powerful opponent, a former pupil of Dumas, challenged (1840) Berzelius to produce for him a single free radical; then he, Laurent, would concede on grounds of analogy the existence of all other presumptive radicals. Alas, proponents of the theory had no reply to make to this challenge, and the theory of radicals might not have been carried over into the fifth decade of the century were it not for the timely appearance of the classic investigations by Bunsen, from the period 1837–42.

The Cacodyl Radical

It had been known since 1760 that when potassium acetate and white arsenic are heated together a fuming offensive smelling substance is produced which is very poisonous and which inflames spontaneously on exposure to air. Bunsen, just starting on his scientific career, made a complete analysis of this so-called "Cadet's fuming liquid." By fractional distillation he separated from it a substance the composition of which he proved to be $((C_2H_6)As)_2O$, for which Berzelius suggested the name "cacodyl oxide." From this oxide a whole series of derivatives was prepared—the chloride, bromide, cyanide, sulfide, etc.—and they all retained the complex (C_2H_6As) unchanged. The original oxide behaved thus like a metallic oxide, with the complex $(C_2H_6)As$ corresponding to the metal. Finally, by treating the cacodyl chloride with zinc, the cacodyl itself was liberated as such. The independent existence of a ternary complex, possessing all the attributes of a metal, was thus no longer fiction. This complex was, in the words of Bunsen, "a true organic element." It is not difficult to comprehend how elated Berzelius was at this discovery. Here at last, coming twenty-five years after cyanogen, was an example of the reality of compound radicals. "This discovery is truly a *curus triumphalis* which will march right through, and will completely smash, Dumas' barricades of bizarre speculations."

Alas, magnificent as was Bunsen's contribution, it could not, standing as an isolated example of free radicals, stem the rising tide of objections to the artificial system of radicals. About this time Liebig was beginning to withdraw from theoretical organic chemistry into agricultural and physiological branches. Berzelius himself was now recognizing clearly the importance of the substitution theories, but in order to uphold at all costs his dualistic system of matter, he was becoming more and more involved in inventing new and strange formulas. Dumas, Laurent, and particularly Gerhard were pushing their new theories through various modifications, which became known as the theory of nuclei, or of residues, or of types, etc. The eclipse of the radical theory seemed imminent, but the situation was saved once more.

The Alcohol Radicals (1849-50)

Bunsen had obtained the cacodyl radical from its chloride by heating the latter with zinc; so now, in 1848, Frankland and Kolbe, in Bunsen's laboratory at Marburg, were trying to obtain the ethyl radical by treating ethyl cyanide with metallic potassium. A reaction did occur, but the result was not the one anticipated: instead of the radical ethyl, they obtained a gas which had the composition (CH_3), and which they considered to be the methyl radical. In order to verify this conclusion, Kolbe undertook to prepare the same methyl by what seemed to him to be a more rational method, by a reaction which would leave no doubt that the methyl radical was actually the one formed. He thus embarked upon his historic experiment—namely, the electrolysis of the potassium salts of the fatty acids. Although the reaction proceeded not quite in accordance with his premises, he found no difficulty in explaining the reaction as it does occur: Acetic acid, considered by him a conjugated compound of methyl with oxalic acid, is broken down by the current into methyl and oxalic acid; the latter becomes at once oxidized to carbon dioxide by the oxygen which results from the simultaneous electrolysis of the water while the hydrogen of the water goes to the cathode. Electrolysis of valeric acid yields similarly the valyl radical.

$$(CH_3).(^1/_2C_2O_3).(^1/_2H_2O) \ = \ \underset{\text{Methyl}}{CH_3 + CO_2} \ + \ H \quad \text{(At anode) (At cathode)}$$

$$(C_4H_9).(^1/_2C_2O_3).(^1/_2H_2O) \ = \ \underset{\text{Valyl}}{C_4H_9 + CO_2} \ + \ H$$

Frankland proceeded in a different manner. At the time of his cacodyl experiments Bunsen, in the hope of obtaining also free alkyl radicals, heated to their respective boiling temperatures various alkyl chlorides with metals. No reaction occurred. "It would," he remarked, "prove of great interest to heat under pressure of their own vapors the various organic chlorides with metals." That is exactly what Frankland undertook to do. Introducing for the first time in the history of chemistry the use of ethyl iodide as a reagent, in preference to the chloride or bromide, Frankland heated under pressure in sealed tubes the iodide with metallic zinc as high as 150° C. On opening the tube he obtained, as one of the several products, a gas of the composition (C_2H_5). He took it to be the free radical ethyl. Similar experiments with methyl iodide and amyl iodide gave corresponding results—the methyl radical and the amyl radical. Summing up, in 1850, his own and Kolbe's results, Frankland concluded: "The isolation of these four compound radicals (methyl, ethyl, valyl, and amyl)—disposes of all doubt as to their actual existence." This conclusion was at once challenged by Gerhardt, by Laurent, by A. W. Hoffmann. They insisted that the substances described should each have their formulas doubled, that they represented in each case, not free radicals, but radical in combination with a similar radical; methyl, for instance, was in reality dimethyl, (CH_3)(CH_3). Kolbe and Frankland took refuge in the reply that, even if so, dimethyl was not necessarily ethane; the former was (CH_3)$_2$, and was a compound radical, while the true ethane was a hydride of the ethyl radical (C_2H_5)H. These two substances, they argued, should not be identical, but merely isomeric, and their own results indicated, in fact, a distinct difference in properties between their dimethyl and ethane. This conclusion was not accepted by many chemists of that time. Nevertheless it remained in the literature undisputed by experimental evidence till 1864, when Schorlemmer definitely proved the identity of dimethyl and ethyl hydride.

Advent of Valence Theory—Structural Chemistry

Thus, with the opening of the second half of the nineteenth century, the theory of radicals as a basis for a comprehensive system of classification received a further reprieve, because once again the actual existence of radicals appeared to be supported by experiment. Meanwhile the whole philosophy of chemistry was undergoing rapid change.

Whether radicals can or cannot exist was becoming a question of secondary importance. Not the composition of the radical but the architecture of the molecule became the all-important consideration. At first only two types of architecture were suggested, their number then increased to four, and the attempt was made to classify all known compounds into the few groups according to the surmised type of their architecture: whether it be that of H_2, H_2O, NH_3, CH_4. (Not to be outdone by the typists, Kolbe proposed his own, and showed that most organic compounds could be considered as derived from carbon dioxide.) Simple types became supplemented by mixed types, and finally the valence theory cemented all the separate parts of truth into one whole. Our present-day organic chemistry came into being. Not the probable function of compacted groups of atoms, not that of radicals, be they real or imaginary, but the function of every individual atom in the molecule—this became the paramount question in chemistry. How well this question has been answered, we all know. The whole dispute as to whether organic radicals can or cannot exist lost completely its original significance. If, as had been postulated by Kekulé, carbon must always function as quadrivalent, then free radicals cannot possibly exist, for they imply bi- or tri-valence of some carbon atom in the molecule. What, then, becomes of those free radicals that have been isolated in the first half of the nineteenth century? A reinvestigation in the light of the new valence hypothesis, supplemented, at this time at the insistence of Cannizzaro, by the forgotten Avogadro hypothesis, discloses the fact that none of these were in reality free radicals. In the light of reinvestigation they now become saturated molecules, the molecular formulas becoming doubled in each case, leaving no carbon atom in the trivalent state. There are no free radicals.

RADICALS (1900—)
Triarylmethyls

For full forty-two years, from 1858 till 1900, the valence hypothesis continued to serve as the one reliable guide in the phenomenal development of organic chemistry. Several successive generations of chemists were brought up with it, and the quadrivalent nature of carbon, the constancy of quadrivalence, served as the cornerstone of this whole development. One by one countless numbers of compounds were assigned a definite constitution or a "structure," to use the happy words of Butlerow (1861).

It was against this background of uninterrupted progress for almost half a century that a publication appeared with the startling title, "Triphenylmethyl, an Instance of Trivalent Carbon."[4] The original intention of the experimenter was to prepare a new substance of the composition $(C_6H_5)_6C_2$, —i. e., hexaphenylethane. The experiment was carried out in strictly orthodox manner, in conformity with well-tried and well-established procedure. The anticipated reaction may be expressed by the following equation:

$$2(C_6H_5)_3C.Cl + 2Ag = (C_6H_5)_3C.C(C_6H_5)_3 + 2AgCl$$

Moreover, in view of the presumptive structure of the new compound, one was justified in predicting the properties that might characterize it. The surprise of the experimenter may well be imagined when the resulting substance possessed not one of the surmised characteristics. Instead of being colorless, it proved to be deeply colored; instead of being stable and inert, it was the very opposite—it instantly combined with the oxygen of the atmosphere. Instead of being unresponsive to the action of iodine, it greedily united with it. It was affected by merest traces of acids; decomposed by light. Instead of being non-electrolytic, it conducted the electric current. In brief, it showed characteristics which singled it out from all known classes of organic compounds. In order to account for this unusual behavior, the plausible hypothesis was finally formulated that the two groups, $(C_6H_5)_3C$, failed to unite with each other, or, if union had occurred, then spontaneous reverse dissociation was also taking place until an equilibrium was established:

$$(R_3C-R_3C) \rightleftharpoons R_3C + CR_3$$

In terms of the old chemistry, we had here free radicals of the order of the methyl; in terms of the modern structural chemistry, it meant that the substance had one carbon atom which functioned, not in the quadrivalent state, but only as *trivalent*.

It was quite natural that, in spite of the substantial arguments advanced in the first publication on this subject, the chemical profession as a whole should show reluctance to accept the views advanced. Suggestions came from various sources, how one might perhaps explain the properties of the

[4] Gomberg, *J. Am. Chem. Soc.*, **22**, 757 (1900).

new substance still in terms of quadrivalence instead of trivalence of carbon. Objections so raised were answered by new and more decisive experiments. After ten years of work in several laboratories in this country and abroad (Chichibabin, Schmidlin, Wieland, Schlenk), all obstacles were swept aside and the doctrine of the constant quadrivalence of carbon was dethroned. By 1910 it became universally recognized that the isolation of free radicals was an accomplished fact. In the course of this long controversy it became necessary to prepare other free radicals, and to study these in detail. At the present time there are known to exist over one hundred substances, each indisputably with a carbon atom in the trivalent state. Trivalency of carbon is no longer anomalous; it is encountered infrequently because the substances of this type are too reactive to persist under our every-day methods of experimentation, and they undergo rapid change into more stable systems.

Stability and Constitution

It has been mentioned above that whenever we speak of free radicals we really have in mind the system indicating equilibrium between free radical and its demolecular normal ethane compound. What are the factors that impart stability to the free radicals, or, in other words, what are the factors that favor dissociation of the ethane molecule into free radicals? In common with all other instances of dissociation, increase of dilution or elevation of the temperature augments the amount of dissociation of the ethane into free radicals. But of far greater importance is the nature of the three groups which are linked to the central carbon atoms. So slight a change as the substitution of $(C_6H_5)_3$ by the group $(CH_3OC_6H_4)_3$ increases the dissociation from 5 per cent to 100 per cent.[5] There are other groups which exert a similar influence, such as naphthyl, biphenyl, etc. Nor is it essential that all three groups linked to the carbon atom be aromatic groups. Through the work of Ziegler, of Conant, of Dufraise, of Schlenk, and of others, we have become familiar with enough instances where one of the three groups may be aliphatic, and still there is dissociation into free radicals. Moreover, at least one of the three groups may be linked to the central carbon atom through oxygen instead of carbon, as the following examples illustrate (R = aromatic group):

[5] Lund, *J. Am. Chem. Soc.*, **49**, 1349 (1927).

R_2C—O—C_6H_5 (Wieland), R_2C—ONa (Schlenk), R_2C—O—MgI (Gomberg and Bachmann).[6] The last two classes, known as metal-ketyls, constitute a very large group, as practically every aromatic ketone is capable of such derivatives. It may also be added that individual compounds have been recently described which are assumed to contain, simultaneously, in one and the same molecule, two carbon atoms each in the trivalent state.[7] One wonders—will this view be ultimately extended to our well-known ethylene compounds, to the benzene molecule?

Anomalous Valence in Elements Other than Carbon[8]

After it had been definitely established that our orthodox conception of valence constancy of carbon had been erroneous, a skeptical attitude was engendered in regard to other elements. It has, in fact, proved possible to prepare radicals wherein elements other than carbon function in anomalous state of valence. In nearly all cases, tin excepting, the element under investigation was loaded to its limit with phenyl or other aromatic groups, since that configuration had proved so successful in the case of the carbon atom. Radicals are at present known with nitrogen bivalent (Wieland, Goldschmidt) instead of the usual trivalent; with sulfur (Lecher) or oxygen (Pummerer, Goldschmidt) as univalent; tin trivalent (Rugheimer, Kraus), and lead as trivalent (Krause); with silicon also presumably trivalent; only recently radicals were described with boron as quadrivalent,[9] and with chromium as quadrivalent and univalent.[10] Again, if results obtained recently in our laboratory (Gomberg and Bachmann) will be confirmed by further work, we shall be in a position to add to that list examples wherein magnesium functions as a univalent element. Using entirely distinct methods, Manchot and his co-workers quite recently found that still other metals possess a latent tendency towards anomalous valence; he speaks of univalent cobalt, nickel,

[6] *Ibid.*, **49**, 236, 2584 (1927).

[7] *Ibid.*, p. 247; Ingold and Marshall, *J. Chem. Soc. (London)*, **129**, 3080 (1926); Scholl, *Ber.*, **60**, 1236 (1927).

[8] Reviews, with references to literature: (a) Schmidlin, "Triphenylmethyl," Enke (1914); (b) Walden, "Chemie der freien Radicale," Hirzel (1924); (c) Gomberg, *Chem. Rev.*, **1**, 91 (1924); (d) Stewart, "Recent Advances in Organic Chemistry," Vol. II, p. 277, Longmans, Green and Co., 1925.

[9] Wahl, *Z. anorg. allgem. Chem.*, **32**, 157 (1926).

[10] Hein and Eissner, *Ber.*, **59**, 362 (1926).

iron, manganese.[11] Barbieri[12] describes compounds of trivalent manganese and of divalent silver. We may also mention the attempts to isolate free acid radicals: CNS^{13a} and ClO_4.[13b]

Time does not permit more than mere mention of the striking spectroscopic evidence in regard to the variability of valence in metals. In the production of their spare spectra, atoms of multivalent metals seem to become stripped of their valence electrons, one by one, giving rise to ion-atom of most varied anomalous valence, such as univalent magnesium, calcium, strontium; univalent and bivalent aluminum, etc. (Fowler; Saunders and Russell). Moreover, there is spectroscopic evidence for the formation of electrically neutral molecules of the type CaCl, when the halide salts of the alkaline-earth metals are fed into a Bunsen flame (Mulliken).

The question arises—shall the name "radical" be applied to every newly discovered compound in which some element functions in a state of valence hitherto considered anomalous for it? Yes and no; it is wholly a matter of definition. (MgI) is as much a radical in its relation to MgI_2 as $(C_6H_5)_3C$ is in relation to $(C_6H_5)_3C.I$. Both are "odd molecules," in the language of G. N. Lewis; each contains an unshared valence electron.

Explaining Mechanism of Reactions

In the study of phenomena three questions present themselves in succession. What happens? How does it happen? Why does it happen? In organic chemistry we know the answer to the first question in a number of cases; but our replies to the second are by no means plentiful.[14]

It seems, here is where the new knowledge in regard to anomalous valence may prove of service. Kekulé, in his pardonable aversion toward all and any radicals, sought to explain the mechanism of chemical reactions by assuming that there is first addition of the reactants to each other, with subsequent split of the complex into two new molecules:

$$(ab) + (a'b') \rightarrow (aba'b') \rightarrow (ab') + (a'b)$$

It is, however, evidently becoming more and more customary to interpret the mechanism of many reactions by assuming

[11] Manchot and Gall, *Ibid.*, **60**, 191, 2175, 2318 (1927).

[12] *Ber.*, **60**, 2421, 2424 (1927).

[13] (a) Soderbäck, *Ann.*, **419**, 219 (1919); (b) Gomberg, *J. Am. Chem. Soc.*, **45**, 398 (1923).

[14] Stewart, "Recent Advances in Organic Chemistry," Vol. II, p. 363.

that the reacting molecules first split into respective radicals, and these then unite to form new molecules. Levine[15] attempts to interpret in this manner the fermentation of glucose. He attributes to the yeast cell the ability to discriminate between glucose passive and glucose active. The active glucose is a free radical, and so falls victim to the yeast cell. Zelinski[16] blames the hexane compound,

$$\begin{array}{c} \overset{|}{C}H \\ H_2C \diagup \quad \diagdown CH_2 \\ | \qquad\qquad | \\ H_2C \diagdown \quad \diagup CH_2 \\ \overset{|}{C}H \end{array}$$

with its two trivalent carbon atoms, for the terrific explosions he has been encountering in the course of his recent investigation. Scheibler[17] claims to have isolated the bivalent radical

$$\mathord{>}\!\!C\!\!\genfrac{}{}{0pt}{}{\diagup ONa}{\diagdown OC_2H_5}$$

and he attributes the formation of methanol from carbon monoxide to intermediate formation of the divalent radical $\mathord{>}\!C\!\genfrac{}{}{0pt}{}{\diagup O\!-\!H}{\diagdown H}$. These are but a few illustrations of the many from the literature of the last few months.

Conclusion

In the evolution of our organic chemical theories for 1800–1860, there was the ever-recurring refrain: Are there radicals; do they have an independent existence? The existence of radicals—i. e., of part-molecules in distinct contrast to whole molecules—has been, in turn, surmised, believed in, considered as demonstrated; then the question became debatable; and finally, beginning with 1860, the independent existence of radicals was looked upon as wholly improbable. After a lapse of half a century the interest in the subject was revived, and the historic quest for the free radicals has been

[15] *Science*, **66**, 560 (1927).
[16] *Ber.*, **60**, 1107 (1927).
[17] *Z. angew. Chem.*, **40**, 1072 (1927).

brought to a successful issue. Hitherto, the definitely recognized units in chemistry have been: atoms, molecules, ions, and electrons. And now, in addition to these four, chemistry, it seems, has to take into account a fifth entity—free radicals.

Bei Fragen zur Produktsicherheit wenden Sie sich bitte an:
If you have any questions regarding product safety,
please contact:

Walter de Gruyter GmbH
Genthiner Straße 13
10785 Berlin
productsafety@degruyterbrill.com